Guide to Growing Strawberries and Berries

Learn the Best Practices for Growing Strawberries and Berries

A. Duller

Lisa Shardon

Copyright © 2024

Guide to Growing Strawberries and Berries

1. Introduction

Importance of Strawberries and Berries

Strawberries and berries represent an essential category of foods from both economic and nutritional perspectives. These fruits are widely used in many cultures due to their culinary versatility and numerous health benefits. They are cherished not only for their sweet and aromatic taste but also for their positive impact on human health.

In addition to being enjoyed fresh, strawberries, raspberries, blueberries, blackberries, and currants are extensively used in desserts, jams, ice creams, and beverages. The growing demand for healthy, antioxidant-rich foods has further increased the value of these fruits, which are regarded as true superfoods. Their economic importance extends beyond national production, with significant exports to European and North American markets, where berry consumption is steadily increasing.

Strawberries and berries also represent a sustainable choice in agriculture. With the right cultivation techniques and crop rotation, they can help preserve biodiversity and reduce environmental impact. Moreover, their use in cosmetics and nutraceuticals—thanks to their beneficial compounds—highlights their strategic importance in wellness and personal care sectors.

Nutritional and Culinary Benefits

The nutritional value of strawberries and berries makes them essential for a balanced diet. They are rich in vitamins, particularly vitamin C and B-group vitamins, as well as minerals like potassium, magnesium, and iron. They also contain antioxidant phytochemicals, such as flavonoids, anthocyanins, and polyphenols, which combat free radicals, slow cellular aging, and reduce the risk of cardiovascular and degenerative diseases.

Regular consumption of these fruits can improve heart health, regulate blood pressure, and enhance brain function, thanks to their antioxidants that fight oxidative stress. They are also a good source of dietary fiber, promoting digestion and satiety, making them ideal for low-calorie diets and weight control.

From a culinary perspective, strawberries and berries are incredibly versatile. They can be eaten plain or incorporated into more elaborate dishes like cakes, sorbets, jams, and juices. Their blend of sweetness and acidity makes them suitable for both sweet and savory preparations. They are often used to garnish salads, yogurt, pancakes, or dishes with fish or fresh cheeses. This wide range of uses makes them valuable ingredients for any season or occasion.

Chapter 1: Types of Strawberries and Berries

Strawberry Varieties

There are many strawberry varieties, each characterized by specific traits regarding flavor, texture, and ripening time. Some of the most well-known varieties include:

1. **Albion Strawberry**

   - **Origin**: United States

   - **Characteristics**: Known for its large, deep-red berries with a bright sheen. Its flavor is sweet and aromatic, with a slight tartness.

   - **Seasonality**: A remontant variety, producing fruit from spring to fall.

2. **Camarosa Strawberry**

   - **Origin**: California

   - **Characteristics**: One of the most widely cultivated commercial varieties. It offers large, cone-shaped fruits with firm texture. It has a balanced taste with good sweetness and acidity.

   - **Seasonality**: Spring-summer.

3. **Mara des Bois Strawberry**

   - **Origin**: France

   - **Characteristics**: This remontant variety is known for its intense wild strawberry aroma and juicy, sweet flesh.

   - **Seasonality**: From spring to autumn.

4. **Elsanta Strawberry**

   - **Origin**: Netherlands

   - **Characteristics**: Among the most popular in Europe, it offers medium-sized fruits with a sweet taste and a pronounced acidity. Its firm texture makes it ideal for transportation and storage.

   - **Seasonality**: May-June.

Each variety has specific climatic and agronomic requirements, making it crucial to select the most suitable type based on local environmental conditions and intended use (fresh consumption or processing).

Berry Varieties

In addition to strawberries, berries include a broad range of small fruits such as raspberries, blueberries, blackberries, and currants, each with distinctive characteristics.

1. **Raspberries (Rubus idaeus)**

   - **Color**: Bright red (also golden in some varieties).

   - **Flavor**: Sweet with slight tartness.

   - **Nutritional Value**: Rich in vitamin C, anthocyanins, and fiber. They have anti-inflammatory properties and promote intestinal health.

- **Varieties**:

  - **Autumn Bliss**: Remontant, producing fruit from August to October.

  - **Tulameen**: Summer-bearing, known for sweet, aromatic fruit.

2. **Blueberries (Vaccinium spp.)**

   - **Color**: Dark blue to black with violet undertones.

   - **Flavor**: Sweet with a hint of tartness.

   - **Nutritional Value**: High in antioxidants, especially anthocyanins, which improve circulation and memory.

   - **Varieties**:

     - **Highbush Blueberry**: Common in North America, producing large, juicy fruits.

     - **Lowbush Blueberry**: Smaller but intensely aromatic berries.

3. **Blackberries (Rubus fruticosus)**

- **Color**: Glossy black.

  - **Flavor**: Sweet with a slight acidity.

  - **Nutritional Value**: Rich in fiber, vitamin C, and antioxidants, supporting cardiovascular health and detoxification.

  - **Varieties**:

    - **Loch Ness**: Large, juicy fruits with no thorns.

    - **Chester**: Cold-hardy, producing fruit from August to September.

4. **Currants (Ribes spp.)**

  - **Color**: Red, white, or black, depending on the variety.

  - **Flavor**: Tart and refreshing.

  - **Nutritional Value**: An excellent source of vitamin C and antioxidants, with diuretic and detoxifying properties.

  - **Varieties**:

    - **Ben Hope Black Currant**: A vigorous, disease-resistant variety.

- **Jonkheer van Tets Red Currant**: Produces sweet and tart clusters in early summer.

Characteristics and Variety Selection

Choosing the right variety of strawberries and berries depends on several factors, including climate, soil conditions, water availability, and intended use. For example, some strawberry varieties are better suited to temperate zones, while others can tolerate hotter or colder climates. Similarly, blueberries require acidic, well-drained soils, while blackberries and raspberries thrive in neutral, fertile soil.

From a commercial standpoint, it is essential to select varieties that ensure high yields and long shelf life. Strawberries like **Elsanta** or **Highbush Blueberries** are highly valued for their transportability and shelf stability. On the other hand, more delicate varieties like **Mara des Bois** or **Autumn Bliss Raspberries** are perfect for fresh consumption and direct-to-consumer sales but less suitable for large-scale export.

For organic farming, it is crucial to prioritize disease- and pest-resistant varieties, reducing the need for chemical treatments. For instance, the black currant **Ben Hope** is known for its natural resistance to various diseases.

Strawberries and berries not only enrich the diet with essential nutrients but also offer endless culinary possibilities. The wide range of available varieties allows growers to meet different production and consumption needs, contributing to the success of these small fruits both in local and international markets.

Chapter 2. Preparing the Soil for Strawberries and Berries

Growing strawberries and berries requires careful soil preparation, as the quality of the soil directly affects plant health and productivity. Each species has specific needs in terms of pH, structure, and soil fertility, but several general practices are essential to ensure an environment conducive to optimal growth. This guide will cover all aspects related to selecting the ideal soil, conducting soil analysis, and preparing and enriching the land.

Choosing the Ideal Soil

1. Sun Exposure and Climate Conditions

The first step in soil preparation is choosing the right growing area, ensuring it provides optimal climatic and environmental conditions. Strawberries and berries thrive in areas with good sun exposure but also require protection from strong winds. Sunlight is particularly important for enhancing the flavor and sugar content of the fruits. However, partial shading may help reduce water stress, especially in areas with very hot summers.

- **Strawberries**: Require at least 6–8 hours of sunlight daily.

- **Blueberries**: Can tolerate partial shade but yield better with full sun.

- **Raspberries and Blackberries**: Prefer full sun but can adapt to partial shade.

- **Currants**: Perform well in cooler areas and tolerate shade better than other berries.

2. Soil Structure and Drainage

Good drainage is essential to avoid waterlogging, which can lead to root diseases such as root rot. Strawberries and berries are sensitive to excess water, as their roots need oxygen to absorb nutrients effectively.

- **Light and Sandy Soil**: Ideal for strawberries, ensuring good drainage and quick warming in spring.

- **Clay Soil**: Unsuitable without structural improvements, as it retains too much water.

- **Acidic, Peaty Soil**: Perfect for blueberries, which require a pH between 4.5 and 5.5.

- **Fertile and Well-Aerated Soil**: Suitable for raspberries, blackberries, and currants.

Soil Analysis

1. Importance of Soil Testing

Before planting strawberries or berries, it is essential to perform soil testing to determine the pH, nutrient content, and physical structure of the soil. This analysis identifies deficiencies in macro- and micronutrients and helps decide if corrective measures are needed. It can also detect issues such as salinity or contamination, which could hinder plant growth.

2. Collecting Soil Samples

Accurate sampling is necessary for reliable results:

- Collect samples from 5–10 different points in the field and mix them to create a representative sample.

- Take samples from a depth of 20–30 cm (the root zone).

- Avoid areas near buildings, ditches, or waterlogged zones.

3. Key Parameters to Analyze

Soil analysis should provide information on the following:

- **Soil pH**:

  - **Strawberries**: 5.5–6.5

  - **Blueberries**: 4.5–5.5

  - **Raspberries and Blackberries**: 5.5–6.5

- **Organic Matter Content**: Determine if

compost or manure is needed to improve fertility.

- **Macro- and Micronutrients**:

  - **Nitrogen (N)**: Essential for vegetative growth.

  - **Phosphorus (P)**: Promotes root development.

  - **Potassium (K)**: Affects fruit quality.

  - **Calcium, Magnesium, and Iron**: Important for plant health.

Preparing and Enriching the Soil

1. Preliminary Soil Preparation

Once the area has been selected and the soil analyzed, the land must be prepared for planting. Soil preparation improves aeration, removes weeds, and facilitates root penetration. Key steps include:

- **Plowing or Digging**:

  - Deep plowing (30–40 cm) improves aeration and breaks up compact layers.

  - In smaller gardens, deep digging is sufficient to loosen the soil.

- **Harrowing or Rototilling**:

  - After plowing, harrowing levels the soil and breaks clods.

  - This step also prepares the soil for mulching or plastic sheeting.

2. Soil Enrichment

To ensure high yields, the soil must be enriched with organic matter and fertilizers according to the results of the soil test.

- **Compost and Aged Manure**:

  - Adding compost or well-decomposed manure during soil preparation enhances fertility and soil structure.

- Recommended dosage: 2–3 kg/m² of compost or manure.

- **Organic and Mineral Fertilizers**:

  - Supplement with nitrogen, phosphorus, and potassium (NPK) fertilizers based on crop needs.

  - **Strawberries**: Require a constant nitrogen supply for fruit production.

  - **Blueberries**: Prefer acidic fertilizers, such as ammonium sulfate.

3. Adjusting Soil pH

If the soil pH is not within the optimal range, it must be corrected before planting.

- **Alkaline Soil**:

  - Add elemental sulfur or acidic peat to lower the pH, especially important for blueberries.

- **Acidic Soil**:

  - Incorporate agricultural lime (calcium carbonate) to raise the pH, particularly for strawberries and raspberries.

4. Weed Control

Weeds compete with crops for light, water, and nutrients, so controlling them is crucial before planting.

- **Mulching**: Apply plastic mulch or organic materials (like straw) to suppress weeds and retain soil moisture.

- **Manual Weeding**: In small fields, remove weeds by hand before planting.

- **Chemical Weed Control**: In large plantations, selective herbicides may be used, though sustainable practices recommend limiting chemical use.

5. Irrigation and Drainage Systems

Proper irrigation is essential for plant growth. Drip irrigation is ideal, as it delivers water directly to the roots, reducing waste and minimizing the risk of foliar diseases.

- **Strawberries**: Require frequent but light watering, as they have shallow roots.

- **Blueberries**: Need consistent moisture without waterlogging.

- **Raspberries and Blackberries**: Benefit from abundant watering during flowering and fruiting.

Preparing the soil for strawberries and berries requires careful planning and targeted actions. Choosing suitable soil, enriching it with organic and mineral fertilizers, managing pH levels, and controlling weeds are essential to ensure successful cultivation. With proper preparation, plants will have the ideal conditions to grow healthily and produce high-quality fruits, meeting both market and consumer demands.

Chapter 3: Caring for Strawberry and Berry Plants

Proper care is essential to ensure healthy growth and high yields for strawberries and berries. Key agricultural practices include appropriate irrigation and drainage management, a balanced fertilization plan, and pest and disease control through natural methods or, when necessary, chemical treatments. In this guide, we will explore the techniques required to maintain plant health and minimize crop losses.

Irrigation and Drainage

Water is a critical factor in the growth of strawberries and berries. However, both overwatering and poor drainage can cause waterlogging and root diseases. Each species has specific irrigation needs that must be met to ensure high-quality fruit production.

1. Strawberries

Strawberries have shallow root systems, making them particularly sensitive to both drought and waterlogging.

- **Irrigation frequency**:

  - Water frequently with small amounts to keep the soil consistently moist without becoming waterlogged.

  - Increase irrigation frequency during flowering and fruiting phases.

- **Irrigation techniques**:

  - **Drip irrigation** is recommended to avoid wetting the leaves, which can promote fungal diseases.

  - Micro-irrigation systems can also be used, though direct overhead watering should be avoided.

- **Drainage**:

  - Well-drained soil is crucial to prevent **root rot**. Raised beds and mulching help maintain proper moisture levels.

2. Berry Plants

Each type of berry has specific water needs:

- **Raspberries and Blackberries**:

  - Require abundant watering during flowering and fruiting. Drip irrigation is ideal for keeping roots moist and preventing fruit rot.

- **Blueberries**:

  - Need plenty of water but no waterlogging. They thrive in **acidic, moist soils**. Mulching with pine needles helps retain moisture and maintain soil acidity.

- **Currants**:

- Require a good supply of water, especially during flowering and fruit swelling. Regular irrigation is essential to prevent water stress that could affect fruit quality.

Fertilization

Balanced nutrition is essential for high yields and healthy fruit. Strawberry and berry plants need **macronutrients and micronutrients** such as nitrogen (N), phosphorus (P), potassium (K), calcium, iron, and magnesium.

1. Strawberry Fertilization

Strawberries have an intensive production cycle and require proper fertilization.

- **Before planting**:
  - Incorporate compost or well-rotted manure

into the soil to improve structure and fertility.

  - Add a **base fertilizer** containing nitrogen, phosphorus, and potassium (NPK), especially before transplanting.

- **During the growing season**:

  - Use slow-release fertilizers to supply nitrogen and potassium throughout the season.

  - Apply foliar fertilizers with micronutrients (such as iron and magnesium) to correct deficiencies.

2. Fertilization for Berry Plants

- **Raspberries and Blackberries**:

  - Apply nitrogen-rich fertilizers early in the season to promote vegetative growth.

  - During fruiting, supplement with potassium to improve fruit quality.

- **Blueberries**:

  - Use **acidifying fertilizers** like ammonium sulfate to maintain low soil pH. Avoid alkaline fertilizers that could damage the roots.

- **Currants**:

  - Apply balanced NPK fertilizers at the start of the season. Add phosphorus to stimulate flowering and potassium during fruit maturation.

Pest and Disease Control

Effective pest and disease control is crucial for keeping plants healthy. Strawberries and berries are susceptible to various diseases and insect attacks, requiring timely prevention and treatment.

1. Common Pests Identification

- **Strawberries**:

  - **Aphids**: Cause leaf yellowing and transmit viruses.

  - **Spider mites**: Lead to discoloration and weaken the plant.

  - **Thrips**: Damage flowers and fruits, causing deformities.

- **Raspberries and Blackberries**:

  - **Spotted wing drosophila (Drosophila suzukii)**: Lays eggs inside fruits, making them inedible.

  - **Beetles**: Attack flowers and fruits.

- **Blueberries and Currants**:

  - **Scale insects**: Feed on plant sap, weakening the plants.

- **Codling moth**: Infests fruits with larvae.

2. Prevention and Natural Treatments

- **Crop rotation**: Reduces the risk of soil-borne diseases and limits pest populations.

- **Mulching**: Suppresses weeds and hinders the development of some insects.

- **Beneficial insects**: Introducing ladybugs and nematodes can help control aphids and other pests.

- **Natural extracts**: Spraying with nettle or garlic solutions can repel pests and strengthen plant resistance.

3. Use of Pesticides and Bioprotection

- **Limited use of chemical pesticides**:

- Use low-impact pesticides only when necessary, following pre-harvest intervals.

- Avoid treatments during flowering to protect pollinators.

- **Bioprotection**:

  - Use **biological products** like *Bacillus thuringiensis* to control specific insects.

  - Set pheromone traps to monitor the presence of adult insects, such as *Drosophila suzukii*.

Caring for strawberry and berry plants requires an integrated approach involving careful irrigation management, a balanced fertilization plan, and effective pest and disease control. Prevention and natural treatments are essential to maintaining plant health while minimizing chemical use. With proper agricultural practices, it is possible to achieve high yields and high-quality crops, ensuring sustainability and respect for the environment.

Chapter 4. Harvesting and Preservation of Strawberries and Berries

Harvesting and preserving strawberries and berries (raspberries, blueberries, blackberries, and currants) are essential steps to ensure product quality while preserving freshness and nutritional value. These fruits are highly delicate and perishable, making it crucial to adopt proper harvesting and storage techniques to minimize damage, maintain organoleptic properties, and extend shelf life.

Harvest Timing

Selecting the right time to harvest strawberries and berries is crucial for obtaining fruits at peak ripeness, flavor, and aroma. Harvesting too early results in unripe, less sweet fruits, while harvesting too late may yield overly soft and perishable produce.

Strawberries

- **Harvest Period**:

  - Strawberries are primarily harvested between **May and July**, though everbearing varieties allow for an extended harvest through the autumn.

  - They must be harvested when they are **fully red and ripe**, as they do not continue to ripen after being picked.

- **Best Times for Harvesting**:

  - Harvest early in the morning or late in the afternoon when temperatures are cooler to reduce stress on the fruits and maintain their freshness.

Raspberries

- Raspberries should be picked when the fruit is **soft and detaches easily** from the receptacle, indicating full ripeness.

- The harvest period ranges from **June to September**, depending on whether the variety is single-crop or everbearing.

Blueberries

- Blueberries are harvested **gradually** since not all the fruits on a cluster ripen at the same time.

- Harvest typically occurs from **July to September**, and fruits should be firm and uniformly blue-black in color.

Blackberries

- Blackberries must be harvested when they are **fully black and soft to the touch**.

- The harvest period extends from **July to September**, and like raspberries, they need to be handled with care to avoid crushing.

Currants

- Red, black, and white currants are harvested between **July and August**, once the clusters are full and the fruits display their characteristic color.

- It is advisable to harvest entire clusters to prevent juice loss from individual berries.

Harvesting Techniques

Since strawberries and berries are fragile, specific techniques must be employed to minimize losses and ensure optimal quality.

Strawberry Harvesting

- **By Hand**:

  - Manual harvesting is the most common method, allowing careful selection of ripe fruits.

  - Strawberries should be picked with their **stem attached**, applying minimal force to avoid damaging the plant.

  - Place the strawberries gently into containers to avoid compressing them.

- **Harvesting Tools**:

  - For larger plantations, **fruit-picking carts** can be used to make the process more efficient and reduce harvesting time.

Berry Harvesting

- **Raspberries and Blackberries**:

  - These berries can be picked by **hand** or with **harvesting combs**, but care must be taken to avoid crushing them.

  - New shoots that will bear the following season's crop should not be damaged during harvest.

- **Blueberries**:

  - They can be harvested manually or with the help of **harvesting combs** for faster picking, especially on large farms.

  - Only fully ripe fruits should be picked, as unripe ones will not continue to ripen after being harvested.

- **Currants**:

  - Currants are usually harvested with the **entire cluster** to prevent berries from breaking. **Scissors** or similar tools can be used to cut the clusters without damaging the plant.

Storage and Post-Harvest Management

Once harvested, strawberries and berries require careful handling to maintain freshness and reduce post-harvest losses. Due to their **perishability**, proper storage practices are essential.

1. Pre-Cooling

- Immediately after harvesting, fruits should be quickly cooled through **pre-cooling** to slow down respiration and the degradation process.

- Ideal pre-cooling temperatures:

  - **Strawberries**: 0–2°C (32–36°F)

  - **Blueberries, Raspberries, Blackberries**: 0–4°C (32–39°F)

  - **Currants**: 0–2°C (32–36°F)

2. Storage in Cold Rooms

- **Cold storage** is crucial to extend the shelf life of these fruits:

  - **Strawberries**: 5–7 days at 0–2°C with 90–95% relative humidity.

  - **Raspberries and Blackberries**: 2–3 days at 0–2°C.

  - **Blueberries**: 10–14 days at 0–2°C.

  - **Currants**: 10 days at 0–2°C.

- Maintaining **high humidity** is essential to prevent the fruits from losing firmness and freshness.

3. Packaging and Transport

- Fruits should be packaged in **small, ventilated containers** to prevent crushing during transport.

- The use of **biodegradable or recyclable packaging** is recommended to comply with environmental regulations.

- During transport, the temperature should remain constant to prevent spoilage.

4. Long-Term Preservation

- **Freezing**:

  - Strawberries, raspberries, blueberries, and currants can be frozen to extend shelf life. It is advisable to freeze the fruits on trays before transferring them to bags to prevent clumping.

  - Frozen berries can be stored for **6–12 months** without significant loss of nutritional value.

- **Drying**:

- Some berries, such as blueberries and currants, can be dried for long-term storage.

- Drying can be done using **electric dehydrators** or sun drying, ensuring the fruits are thoroughly cleaned and free of moisture.

- **Making Preserves and Jams**:

  - Strawberries and berries can be processed into **jams, preserves, juices, and syrups** to increase their shelf life and add value.

5. Post-Harvest Use

- **Fresh Markets**: Fresh fruits are sold directly to consumers through local markets and specialty stores.

- **Food Industry**: Some production is destined for the food industry for making yogurt, desserts, and beverages.

- **Gastronomy**: Berries are essential ingredients in desserts, cakes, ice cream, and cocktails.

The harvesting and preservation of strawberries and berries require meticulous planning and careful management to ensure product quality and freshness. Harvesting must be done at the right time and with appropriate techniques to minimize damage and losses. Cold storage, freezing, and drying help extend shelf life, while value-added products like preserves contribute to maximizing the fruits' potential. Proper post-harvest handling is crucial not only to retain the nutritional and organoleptic properties but also to meet market demands with high-quality produce.

Chapter 5. An Overview of Growing Strawberries and Berries in Pots

Growing strawberries and berries in pots is a valid alternative to traditional open-field cultivation. This method offers several advantages, such as optimizing limited spaces and better controlling environmental conditions. Strawberries, raspberries, blueberries, blackberries, and currants adapt well to container cultivation, provided that their needs for watering, exposure, and nutrition are properly met. This guide explores the **benefits of container gardening**, how to **select suitable pots**, and the **care techniques** required to ensure healthy and productive plants.

Benefits of Growing in Pots

Container gardening overcomes some of the limitations of traditional methods and offers many advantages, especially for those with limited outdoor space, such as balconies, terraces, or courtyards.

1. Space Optimization

- Container gardening is ideal for people living in cities or without access to a garden.

- Pots can be placed on **balconies, terraces, courtyards**, or windowsills, making the most of every available space.

- Vertical growing structures, such as **multi-level or stacked planters**, can further increase production capacity.

2. Control of Soil Conditions

- Growing in pots allows for **better control of the substrate**, ensuring fertile and well-drained soil.

- Specific soil mixes can be chosen for each plant type: for example, blueberries require acidic soil, which is easy to achieve with acidophilic potting mixes.

- Potted plants are less prone to soil diseases,

such as root rot, which are common in open-field cultivation.

3. Greater Plant Mobility

- Potted plants can be **moved easily** to find better sun exposure or protect them from wind and frost.

- During winter, containers can be moved to sheltered areas to ensure the survival of more delicate plants.

4. Fewer Issues with Pests and Weeds

- Container gardening significantly reduces the growth of **weeds**, which can be difficult to control in open soil.

- Soil-borne diseases are less likely to develop in containers, reducing the need for pesticides and promoting sustainable cultivation.

Choosing Containers for Strawberries and Berries

Selecting the right containers is crucial to ensure healthy plant growth. Each species has specific requirements for pot depth, volume, and drainage.

1. Types of Pots

- **Plastic Pots**:

  - Lightweight, affordable, and easy to move, though they can overheat in the sun, damaging roots.

- **Terracotta Pots**:

  - Porous and breathable, they promote good soil aeration but tend to dry out faster and are heavier to move.

- **Vertical Planters or Growing Towers**:

  - Useful for growing strawberries and berries on multiple levels, optimizing vertical space.

- **Grow Bags**:

- Ideal for temporary cultivation, providing excellent drainage and easy disposal at the end of the season.

2. Container Size

- **Strawberries**:

  - Require pots with a **minimum depth of 20-30 cm** and good drainage. Long planters or **pocketed vertical containers** are also suitable.

- **Raspberries and Blackberries**:

  - Need larger, deeper pots, at least **40-50 cm**, to allow root development.

- **Blueberries**:

  - Require large, deep containers, **50-60 cm**, with a specific acidophilic soil mix.

- **Currants**:

  - Can be grown in containers **30-40 cm deep** but need space to spread horizontally.

3. Proper Drainage

- All containers must have **drainage holes** to prevent waterlogging, which can lead to root rot.

- Adding a layer of **expanded clay or gravel** at the bottom of the pot helps improve drainage.

Care for Potted Plants

Potted plants require special care, as they have limited access to water and nutrients compared to those grown in open soil. Below are key aspects to consider.

1. Watering

- Potted plants need **more frequent watering** because the soil dries out quickly.

- It's essential to avoid both overwatering and underwatering; the soil should be kept **moist but not waterlogged**.

- Installing **drip irrigation systems** or self-watering reservoirs ensures a consistent water supply, especially in hot weather.

2. Fertilization

- Nutrients in potted soil deplete quickly, so it's necessary to use **slow-release fertilizers** or apply liquid fertilizers regularly.

- For strawberries, it's important to supplement with **potassium and phosphorus** during flowering to encourage fruit development.

- Blueberries require **acidifying fertilizers**, such as ammonium sulfate, to maintain the soil's low pH.

3. Sunlight Exposure

- Strawberries and most berries need **at least 6 hours of direct sunlight** per day for optimal fruit production.

- In regions with particularly hot summers, moving the pots to **partial shade** during the hottest hours can prevent plant stress.

4. Pruning and Plant Management

- **Strawberries**: Remove **stolons (runners)** to prevent energy waste and encourage higher fruit yield.

- **Raspberries and Blackberries**: Prune old and non-productive canes to stimulate new fruit-bearing shoots.

- **Blueberries and Currants**: Annual pruning is necessary to maintain plant health and productivity by removing dry or overly mature branches.

5. Pest and Disease Control

- While potted plants are less susceptible to soil-borne diseases, they can still be affected by **aphids, spider mites, or scale insects**.

- Spraying with **natural macerates**, such as nettle or garlic extract, can help prevent infestations.

- Regular monitoring and quick intervention

are necessary to control fungal infections, such as powdery mildew or gray mold.

6. Winter Preparation

- Some plants, such as raspberries and currants, are cold-hardy, but others, like strawberries and blueberries, may require winter protection.

- During the coldest months, it's advisable to **move pots to sheltered areas** or cover them with **fleece or non-woven fabric** to protect them from frost.

Growing strawberries and berries in pots is a practical and accessible solution for anyone wishing to enjoy fresh fruit without a garden. With the right container selection, proper watering, and balanced nutrition, along with careful plant management, it's possible to achieve abundant and high-quality harvests.

Chapter 6: Recipes and Culinary Uses of Strawberries and Berries

Strawberries and berries, such as raspberries, blueberries, blackberries, and currants, are not only delicious and nutritious but also extremely versatile in the kitchen. These fruits can be used in a wide range of recipes, from sweet to savory dishes, making them an ideal ingredient for any occasion. In this article, we'll explore creative ideas for using strawberries and berries in cooking, along with how to preserve them by preparing jams and syrups.

Ideas for Using Strawberries and Berries

1. Sweets and Desserts

a. Cakes and Tarts

- **Strawberry Tart**:

- Prepare a shortcrust pastry base and fill it with pastry cream. Garnish with halved fresh strawberries and a thin layer of jelly to glaze.

- **Blueberry Cake**:

  - Use a basic butter cake recipe and fold in fresh blueberries. Bake until golden and dust with powdered sugar before serving.

b. Cheesecakes

- **Strawberry Cheesecake**:

  - Create a base from crushed biscuits and butter, and prepare the filling with cream cheese, sugar, and whipped cream. Top it with a fresh strawberry sauce.

- **Berry Cheesecake**:

  - Make a similar version with a blend of raspberries, blueberries, and blackberries in both the filling and topping.

c. Ice Creams and Sorbets

- **Strawberry Ice Cream**:

- Blend fresh strawberries with sugar and a little lemon juice. Add whipped cream and freeze.

- **Berry Sorbet**:

  - Blend raspberries, blueberries, and sugar, then freeze, stirring occasionally for a creamy texture.

d. Muffins and Pancakes

- **Blueberry Muffins**:

  - Add fresh blueberries to the muffin batter for a burst of sweetness.

- **Strawberry Pancakes**:

  - Mix pureed strawberries into the pancake batter and top with fresh strawberries and maple syrup.

2. Savory Dishes

a. Salads

- **Spinach and Strawberry Salad**:

  - Combine fresh spinach, sliced strawberries, pecans, and goat cheese. Dress with a balsamic vinaigrette.

- **Quinoa and Berry Salad**:

  - Mix cooked quinoa with blueberries, raspberries, cucumbers, and crumbled feta. Drizzle with olive oil and lemon juice.

b. Sauces and Condiments

- **Strawberry Sauce**:

  - Blend fresh strawberries with balsamic vinegar, honey, and a pinch of salt to create a sauce for grilled meats.

- **Blackberry Chutney**:

  - Cook blackberries with onions, apple cider vinegar, and spices to make a chutney that pairs well with cheese or meats.

3. Beverages

a. Smoothies and Shakes

- **Strawberry Banana Smoothie**:

  - Blend fresh strawberries, banana, yogurt, and some milk for a creamy, nutritious smoothie.

- **Berry Smoothie**:

  - Combine raspberries, blueberries, Greek yogurt, and orange juice for an antioxidant-rich smoothie.

b. Cocktails

- **Strawberry Mojito**:

  - Mix fresh strawberries, mint, lime, and rum for a refreshing summer cocktail.

- **Berry Bellini**:

  - Blend raspberries and mix with prosecco for a bubbly, fruity cocktail.

4. Preserves and Preparations

a. Jams

- **Strawberry Jam**:

  - Cook fresh strawberries with sugar and lemon juice until thick. Sterilize jars and fill them with the hot jam.

- **Mixed Berry Jam**:

  - Combine raspberries, blueberries, and blackberries with sugar and some pectin to make a flavorful jam.

b. Syrups

- **Strawberry Syrup**:

  - Cook strawberries with sugar and water until a thick syrup forms. Strain and store in glass bottles. Use to sweeten drinks or as a topping for desserts.

- **Berry Syrup**:

  - Simmer mixed berries with sugar and water to make a versatile syrup for cocktails, sweets, and desserts.

Preserving with Jams and Syrups

1. Preparing Jams

Making jam is an excellent way to preserve strawberries and berries, allowing you to enjoy their flavor year-round.

Basic Ingredients

- **Fruits**: Strawberries, raspberries, blueberries, blackberries, currants.

- **Sugar**: The amount of sugar depends on the sweetness of the fruits and the recipe. Usually, a 1:1 ratio of sugar to fruit is used.

- **Lemon Juice**: Helps preserve color and flavor while balancing sweetness.

- **Pectin** (optional): Assists in thickening the jam, especially for fruits with low pectin content, like strawberries.

Steps

1. **Fruit Cleaning**: Wash the fruits

thoroughly, removing stems and leaves.

2. **Fruit Preparation**: Chop the strawberries and lightly crush raspberries and blueberries to release their juices.

3. **Cooking**: In a large pot, combine the fruit, sugar, and lemon juice. Bring to a boil over medium heat, stirring occasionally.

4. **Consistency Check**: Cook until the jam reaches the desired consistency. Test by placing a spoonful on a cold plate; if it doesn't slide, the jam is ready.

5. **Sterilizing Jars**: Sterilize jars in boiling water and let them dry.

6. **Filling the Jars**: Pour the hot jam into sterilized jars, leaving some space at the top. Seal with airtight lids.

7. **Storage**: Let the jars cool to room temperature and store them in a cool, dark place. When properly stored, the jam can last up to a year.

2. Preparing Syrups

Berry syrups are excellent for sweetening

drinks, making desserts, or as toppings for pancakes and ice creams.

Basic Ingredients

- **Fruits**: Strawberries, raspberries, blueberries, blackberries.

- **Sugar**: The amount varies, usually equal to or more than the amount of fruit.

- **Water**: Needed for dilution and syrup preparation.

Steps

1. **Preparing the Fruit**: Wash and cut the fruits.

2. **Cooking**: In a pot, combine the fruits, sugar, and water. Bring to a boil, stirring to dissolve the sugar.

3. **Enhancing the Flavor**: Simmer over medium-low heat for 10-15 minutes until the syrup thickens.

4. **Straining**: Filter the syrup through a

fine mesh sieve or cheesecloth to remove solids.

5. **Bottling**: Pour the hot syrup into sterilized bottles and seal tightly.

6. **Storage**: Syrups can be refrigerated for several weeks or sterilized for long-term storage.

Strawberries and berries offer endless possibilities in the kitchen, allowing you to create tasty, healthy, and colorful dishes. Their versatility makes them suitable for both sweet and savory recipes, making them essential ingredients in every kitchen. Additionally, preserving them through jams and syrups ensures that their flavors can be enjoyed even during the winter months. Experimenting with strawberries and berries not only enriches your table but also opens the door to a world of flavors and combinations.

Glossary of Agricultural Terms: Strawberries and Berries

The cultivation of strawberries and berries is a field rich in specific terminology and agricultural techniques that deserve exploration. This glossary aims to clarify some of the most commonly used terms in this sector, facilitating the understanding of agricultural practices and fundamental concepts related to the cultivation of these delicious fruits. Additionally, at the end of the glossary, there are some reflections on sustainable cultivation, an increasingly important approach in modern agriculture.

Glossary

A

- **Acidity**: Measure of the pH of soil or fruit. Strawberries and berries generally prefer a slightly acidic pH (5.5-6.5).

- **Antioxidants**: Compounds found in fruits and vegetables that help fight free radicals in the human body, contributing to overall health.

- **Apomixis**: Formation of seeds without fertilization. Some varieties of strawberries can reproduce through this process.

B

- **Biotechnology**: Technologies that apply biological principles to improve crops. It can include advanced selection techniques or genetic engineering.

- **Phytosanitary bulletin**: Informative document that provides guidance on the main pests and diseases that can affect crops.

C

- **Suckering**: Growth of new shoots from the base of the plant. This process is common in strawberries, contributing to vegetative propagation.

- **Life cycle**: Stages of development of a plant, from planting to maturation and harvest.

- **Climate**: Average atmospheric conditions of a given area that affect plant growth. Strawberries and berries require specific climates to thrive.

D

- **Drainage**: Removal of excess water from the soil, essential for preventing root rot in strawberry and berry plants.

- **Desiccation**: Process of drying plants, useful in some agricultural practices for pest control.

F

- **Fertilization**: Application of nutrients to the soil to promote plant growth. Strawberries and berries require balanced fertilizers to achieve good results.

- **Phytopathology**: Study of plant diseases, including pathogens that can affect strawberries and berries.

G

- **Germination**: Process in which a seed begins to develop into a new plant. It may require specific temperature and humidity conditions.

- **Grafting**: The process of joining a plant variety onto another to improve its resistance or productivity.

I

- **Irrigation**: Provision of water to plants. It is essential for the growth of strawberries and berries, especially during dry periods.

- **Intercropping**: Agricultural practice of cultivating different plant species in the same field to maximize land use and reduce competition among plants.

L

- **Soil tillage**: Preparation of the land for planting, which may include plowing, tilling, and leveling.

- **Fungal diseases**: Plant diseases caused by fungi, common in strawberry and berry crops.

M

- **Microclimate**: Variability of climatic conditions in a small area. Creating microclimates can encourage plant growth.

- **Mulching**: Technique of covering the soil with organic or inorganic materials to conserve moisture, reduce weeds, and improve soil health.

P

- **Pests**: Organisms that damage plants, such as aphids, mites, and nematodes. Pest control is crucial for the health of strawberries and berries.

- **Dioecious plants**: Plants that have male and female flowers on separate individuals. Some berries, such as currants, may reproduce this way.

- **Propagation**: Method of multiplying plants, which can occur by seed, cutting, or division.

R

- **Harvest**: The process of collecting ripe fruits. Strawberries and berries must be harvested at the right time to ensure maximum flavor and quality.

- **Disease resistance**: The ability of a plant to withstand pathogens and diseases. Some varieties of strawberries and berries have been selected for increased resistance.

S

- **Sustainability**: Agricultural practices aimed at preserving the environment, maintaining biodiversity, and ensuring the health of soil, water, and plants in the long term.

- **Soil**: The upper layer of the Earth that supports plant life. Soil quality is fundamental for the growth of strawberries and berries.

T

- **Transplanting**: Moving seedlings to another location to encourage growth. This practice is common in strawberry cultivation.

- **Phytosanitary treatments**: Preventive or curative measures to protect plants from pests and diseases.

V

- **Variety**: Diversity of a plant species with specific characteristics, such as flavor, color, and disease resistance. There are many varieties of strawberries and berries, each with unique features.

- **Wild vegetation**: Plants that grow naturally in a given environment without human intervention. Wild vegetation can influence the growth of crops.

Final Reflections on Sustainable Cultivation

Sustainable cultivation of strawberries and berries is essential for ensuring a healthy and productive future. This approach involves adopting practices that preserve the environment, enhance biodiversity, and ensure the quality of soil and water.

1. Regenerative Agriculture Practices

Regenerative agriculture focuses on restoring soil health through practices such as crop rotation, composting, and cover cropping. These techniques improve soil structure, increase water retention, and promote microbial biodiversity, which is essential for plant growth.

2. Responsible Resource Use

Irrigation is a critical aspect of strawberry and berry cultivation. Drip irrigation techniques and rainwater harvesting systems can reduce water consumption while ensuring that plants receive the right amount of moisture.

3. Biological Pest Control

Using natural methods for pest control, such as introducing beneficial insects (e.g., ladybugs or parasitic wasps) and using traps, can reduce the need for chemical pesticides, minimizing the impact on the environment and human health.

4. Education and Awareness

Encouraging growers to learn about sustainable practices is crucial for the success of sustainable cultivation. Workshops, training courses, and educational programs can raise awareness and provide farmers with the skills needed to implement more eco-friendly techniques.

5. Collaboration and Networking

Agricultural communities can benefit from collaboration by sharing resources and knowledge. Local networks can facilitate access to updated information and innovative technologies, promoting sustainable agricultural practices.

In conclusion, sustainable cultivation of strawberries and berries not only improves the quality and quantity of production but also contributes to the health of our planet. Investing in sustainable practices is a crucial step toward ensuring prosperous and responsible agriculture capable of facing future challenges.

Index

1. Introduction pg.4

Chapter 1: Types of Strawberries and Berries pg.7

Chapter 2. Preparing the Soil for Strawberries and Berries pg.14

Chapter 3: Caring for Strawberry and Berry Plants pg.23

Chapter 4. Harvesting and Preservation of Strawberries and Berries pg.33

Chapter 5. An Overview of Growing Strawberries and Berries in Pots pg.43

Chapter 6: Recipes and Culinary Uses of Strawberries and Berries pg.52

Glossary of Agricultural Terms: Strawberries and Berries pg.62

www.ingramcontent.com/pod-product-compliance
Lightning Source LLC
Chambersburg PA
CBHW071747240526
45471CB00023B/2670